Plants of the Texas Shore

Yucca along San Antonio Bay

Plants of the Texas Shore

A BEACHCOMBER'S GUIDE

By Mary Michael Cannatella and Rita Emigh Arnold

Drawings by Suzanne Gillespie Daughtery

Photographs by Mary Michael Cannatella

Published with the Texas A&M University Sea Grant College Program
by Texas A&M University Press, College Station

Partially supported through Institutional Grant NA83AA-D-00061 to Texas A&M University by the Office of Sea Grant, National Oceanic and Atmospheric Administration, Department of Commerce. Texas A&M University Publication Number TAMU-SG-84-301.

Library of Congress Cataloging in Publication Data

Cannatella, Mary Michael, 1958–
Plants of the Texas shore.

"Texas A&M University publication number TAMU-SG-84-301"—T.p. verso.
Includes index.
1. Seashore flora—Texas—Identification. 2. Seashore ecology—Texas. I. Arnold, Rita Emigh, 1950–
II. Title.
QK188.C36 1985 582.13'097641 84-40553
ISBN 0-89096-214-6

Manufactured in the United States of America
FIRST EDITION

For my parents

M.M.C.

Be praised, My Lord, for our Sister, Mother Earth,
who holds us up and keeps us straight, yielding diverse
fruits and flowers of different hue, and grass.

Francis of Assisi

Contents

MAP

Acknowledgments

I wish to thank Dr. Clarissa Kimber and my graduate committee members for their assistance in preparing the report upon which this book is based and Dr. Melvin Schroeder for his review. Special thanks go to Colin and Mischelle for their VW and to Marc for his time, patience, and driving lessons.

M.M.C.

Plants of the Texas Shore

THE TEXAS COAST
Sabine River
Sabine Lake
Houston
Chenier Plain
Galveston Bay
Bolivar Peninsula
Galveston
Galveston Island
Lavaca Bay
Matagorda Peninsula
San Antonio Bay
Matagorda Bay
Copano Bay
Matagorda Island
Aransas Bay
Corpus Christi Bay
San Jose Island
Corpus Christi
Mustang Island
Baffin Bay
Gulf of Mexico
Laguna Madre
Padre Island
Rio Grande
0
25
50
75
miles
Brownsville

Introduction

Texas annually welcomes thousands of visitors to her shores, where land meets sea for nearly four hundred miles. The coastal zone teems with life, from tiny ghost crabs scurrying across sandy beaches to five-foot-tall whooping cranes wintering in Texas marshlands. But plants, the basis of life in the dunes and wetlands, often go unnoticed by visitors. Plants provide a background mosaic for the sea and sun that lure people to the coast to fish, swim, hunt, and relax; but more importantly, plants are a vital source of nutrients for animal life.

This publication describes major habitats along the Texas coast and identifies the plants—the foundation of the food web—to be found in each. The plants and animals of the coastal zone are special, having adapted to wind, salt, and sand in an environment that is frequently flooded. They are survivors.

For those willing to explore beyond the hotels, restaurants, and amusement parks that are part of the coastal landscape, a variety of beautiful and interesting plants, both exotic and commonplace, are waiting. We hope this guide will help amateur naturalists, students, coastal visitors, or anyone curious about

natural life along the sea to appreciate the bountiful but fragile life on the Texas shoreline.

People have lived in Texas for the past eleven thousand years, maybe longer. Along the Gulf coast, Native Americans scavenged the waters and lands in search of food. These Indians, now extinct, left a record of their existence in pottery fragments, stone tools, and shell middens (refuse heaps). Archaeologists have been studying these remnants to learn more about the early coastal inhabitants. The artifacts, along with written descriptions by Cabeza de Vaca and other European explorers, present a sketchy view of what life was like for early coastal people.

When Europeans first came to Texas in the sixteenth century, Texas' coastal tribes occupied the entire shore zone, from the Sabine River south to the Rio

Grass at water's edge

Grande. The Atakapas lived along the chenier plain from the Sabine Pass to Galveston. The Karankawas occupied the lands between Galveston and Corpus Christi, and the Coahuiltecans lived between Corpus Christi and the Rio Grande. Each tribe was made up of small, seminomadic bands that hunted game, gathered wild plants, and fished the Gulf, bay, and river waters. The search for food was based on seasons. Spring and summer were spent on the barrier islands and beaches; fall and winter were passed on the mainland. Seeds and rhizomes (underground stems) of the American lotus and the fruits of prickly pear cacti, which we call tunas, were gathered from the wild. Along the southernmost reaches of the coast, the Coahuiltecans ate agave (century plant) flowers as well as prickly pear tunas. The seas provided various fish, shellfish, and turtles for the Indians to eat. Alligators were hunted for meat, and the grease of the reptile was used as an insect repellant. The Karankawas used marine shells as tools and ornaments.

These coastal tribes sought protection from the winter cold on the mainland. There they hunted javelinas and any bison that had migrated to the coastal plain. Deer and rabbits also provided meat as well as skins for capes and blankets for the scantily clad natives. The forests of the riverbanks were a source of pecans, acorns, and seeds. Life was harsh for these early inhabitants, and they counted on the coast and its resources for bare essentials.

Today man still exploits the coast, as did his predecessors, but on a much larger scale. Fishing, still a popular coastal activity, has become a billion-dollar industry. The sale of fish and shellfish—red drum, spotted seatrout, flounder,

Shrimp boat

shrimp, oysters, and crabs—provides a livelihood for thousands of coastal residents. The discovery of oil in 1901 at Spindletop, near Beaumont, ushered in a new major economic factor along the coast, the petrochemical industry.

Besides providing resources for industry, the Texas coast offers recreational opportunities which attract thousands of hunters, photographers, naturalists, and tourists each year. Miles of Texas beaches have been developed into resort areas, and miles more remain pristine. Sections of the coast have been set aside by the state and federal governments as refuge areas for native flora and wildlife, and a variety of birds, reptiles, and other animals depend on vegetation all along the coast for food and sometimes shelter.

What types of wildlife are most common along Texas shores today?

Grass and shells at the edge of a bay

Countless sea gulls and sandpipers frequent the shores off the barrier islands, feeding on fish and small sand-burrowing creatures. Sanderlings, willets, plovers, and skimmers scavenge the shore for crustaceans, marine worms, or tiny mollusks. Terns scamper along the water's edge, leaving tiny imprints in the sand alongside ghost crab tracks.

Small crustaceans also live in this

Gull on the beach

Sea gulls

swash zone, where waves wash up onto the sand. But they cannot be seen because they burrow into the wet sand as they are washed ashore. Remnants of other creatures, however—those that lived offshore, died, and were carried ashore—are visible along barrier island shorelines. They include moonsnail shells, cockle shells, coquina shells, and the sand dollar. On chenier beaches, shells are found in abundance. Eastern murex, coquina, and moonsnail shells are strewn along the beaches paralleling Highway 87 between Galveston and Beaumont.

Hidden away within the barrier islands' massive sand dunes are creatures seeking refuge from the intense Texas sun. Ord's kangaroo rat lives among the dunes of Padre Island, burrowing into the sand by day and feeding on seeds at night. Two other nocturnal rodents,

Sand dollar

the Texas pocket gopher and the ground squirrel, seek shelter in the dunes. Tiny ghost crabs burrow into the same sands, leaving distinctive tunnels in both the dunes and the swash zone. What remains on the surface are tiny, telltale holes that mark their escape routes. Other imprints among sea oats and paspalum grasses trace paths taken by snakes, willets (large shore birds), and sandhill cranes.

The mourning dove, the most popular game bird in the state, inhabits the grassland flats of the barrier islands as well as every county in Texas. The armadillo and opossum frequent the flats, as do various rodents that provide food for the venomous western diamondback rattlesnake. The western coachwhip snake is another predator of rodents that stalks the flats, feeding on insects, lizards, and young birds. Toads and frogs are generally scarce on the barrier islands, and the lack of fresh water on the southernmost barrier islands limits amphibian populations there.

The rich assortment of plants on the chenier ridges and in the marshes among them provides ample roosting, feed-

Alligator in a chenier marsh

ing, and nesting sites for both terrestrial and wetland fauna. Herons nest in trees and tall shrubs near freshwater marshes, ponds, and streams. Pelicans, terns, and sea gulls visit the beach front, skimming the shallow waters for fish and small crustaceans. Chenier marshes also provide habitats for rails, hawks, and coots as well as nutria, raccoons, and white-tailed deer. Various ducks are common, too, as are reptiles such as the cottonmouth snake and the American alligator.

Many animals use Texas' other marshes and estuaries as permanent or temporary quarters to feed, winter, or spawn. Nutria and muskrats feed on clusters of soft stem rushes in fresh marshes and on various parts of arrowhead and bulltongue. Ducks eat stem rushes too, and, along with other marsh birds, enjoy knotweed. In brackish marshes the seeds of saltgrass and marshhay cordgrass are also a favorite of ducks. Nutria, snow geese, and, less often, cattle consume the roots, stems, and leaves of these species as well.

Reptiles and amphibians such as toads, frogs, and an assortment of snakes inhabit Texas marshes, along with mammals like nutria, raccoons, mink, and white-tailed deer. The Kemp's Ridley, leatherback, and green turtles all roam Gulf waters and adjacent bays; Canadian geese, snow geese, ibises, anhingas, and white and brown pelicans are seasonal residents of the coastal plain. Roseate spoonbills, snowy egrets, cattle egrets, and several species of herons can be seen near inland lakes and ponds. The most famous bird is the endangered whooping crane, which annually winters at Aransas National Wildlife Refuge near Matagorda Island.

Most fishes of the Gulf of Mexico off

Heron

Texas spend part of their life cycles in shallow bay waters, part in deep Gulf waters, and part in the low-lying marshes. Fishes not only provide a livelihood for coastal residents but also draw sport fishermen to the coast. Popular catches include spotted seatrout, redfish, flounder, black drum, and catfish. Crabs, oysters, and shrimp are taken by individuals and fished commercially as well.

Man depends on the Texas shore for food, energy, and recreation. It is home to more than one million people, as well as a habitat for the variety of animals mentioned and an array of plants, the most common of which will be described in the following pages.

The Texas coast is a vital part of our heritage and our lives. We need to protect its environments for the flora and fauna as well as for ourselves and future coastal residents.

The Texas Coast

The word *coastline* brings to mind many different images: a crowded beach where children build sand castles and adults sunbathe; condominiums forming a concrete skyline against an otherwise flat terrain; a deserted beach where silence is broken only by winter surf and screeching gulls; or a vast marshland where hunters can bag waterfowl or white-tailed deer. The Texas coast is all these things: undeveloped beaches, Gulfside resorts, acres of wetlands, and wildlife refuges. Even a national seashore is part of the state's coast.

Although the Texas coast lends itself to a variety of uses, the gradually sloping, flat coastal plain has three main components: barrier islands, bays and marshes, and a chenier plain. Each landscape harbors different habitats that accommodate certain plants and animals adapted to the rugged environment.

Texas' sandy barrier islands, muddy bays, extensive dune fields in the south, and biologically productive marshes in the north did not always look the way they do today. The state's shoreline was affected by changes in sea level during what we commonly call the Great Ice Age. At that time, an estimated two to three million years ago, glaciers grew, siphoning the oceans' waters and advancing on the continents. By locking up the water in

Condominiums in the dunes

Padre Island beach

immense ice sheets, the glaciers lowered the sea level and pushed coastlines miles offshore from where they are today. Eventually the glaciers melted and retreated, allowing water to drain back into the oceans and causing the sea level to rise again.

This cycle happened several times during the Great Ice Age. Each time, the position of the shoreline changed. It is believed that as recently as eighteen thousand years ago, Texas beaches were more than fifty miles offshore from their current position, because the sea was almost four hundred feet below its present level.

Although glaciers caused dramatic

Weathered oaks on the shore

changes in the state's shoreline, more subtle changes occur day-to-day in our time. These are most evident on the barrier islands.

Barrier Islands

Barrier islands—long, narrow, gradually sloping islands—are probably the most common feature along the Texas

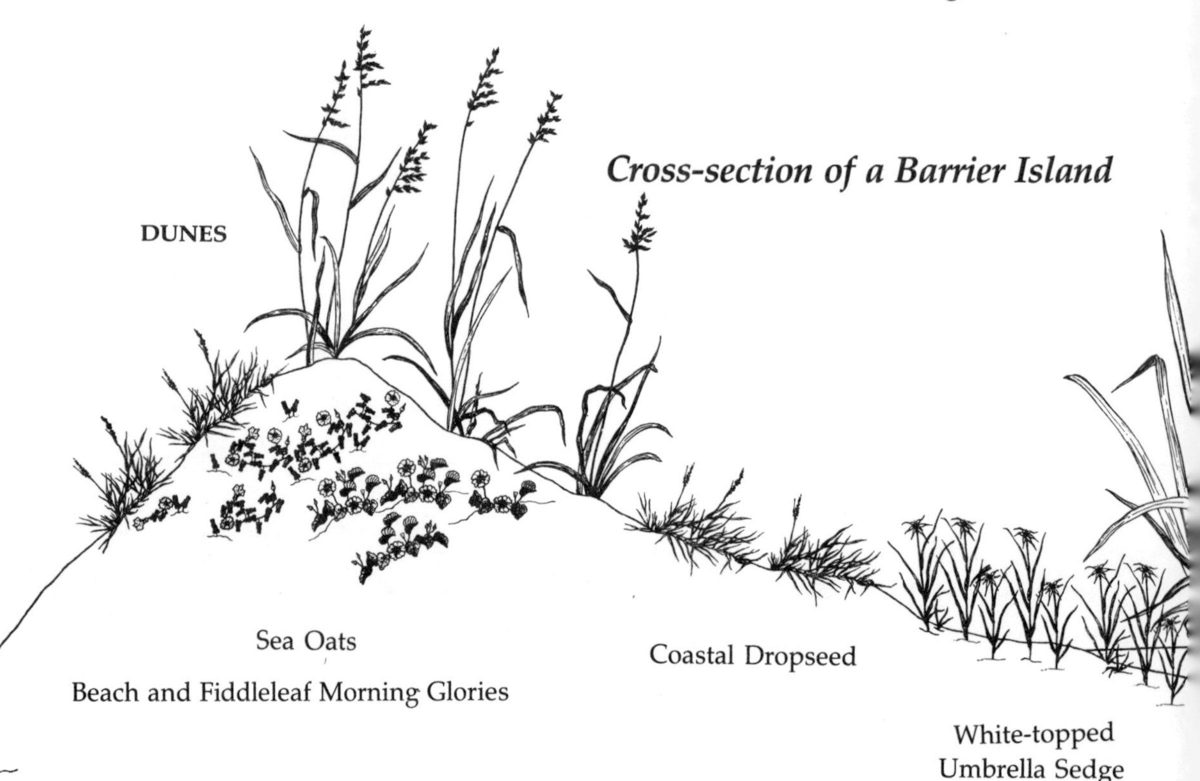

Cross-section of a Barrier Island

coast. Of the nearly 400 miles of coastline, almost 320 miles are fronted by the barrier island chain. The most popular Texas beaches are on that chain: Padre Island, famous as a resort and national seashore, and Galveston and Mustang islands, also noted recreation spots.

Barrier islands are not uncommon coastal features in the United States. The southern Atlantic and northern Gulf of Mexico coasts are all trimmed by these islands that serve as buffer zones, protecting the mainland from frequent tropical storms and hurricanes. They are dynamic; active shore currents constantly deposit and remove sand from them.

Because they are always in a state of flux, barrier islands are fragile areas. Any abrupt change, man-made or otherwise,

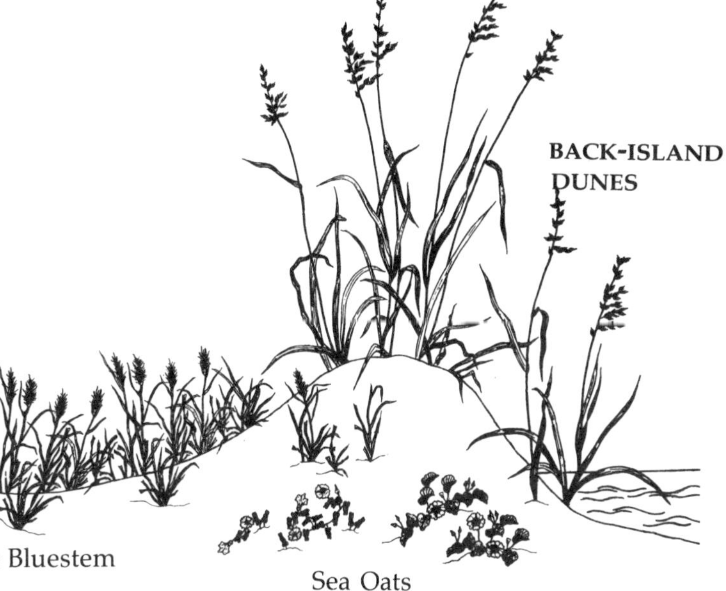

can strongly affect them. A severe hurricane, for example, can breach or break through a weak spot in an island, creating small channels or large fan-shaped sand deposits called washover fans.

Though many different theories have been proposed to explain how barrier islands were formed, geologists believe that the origin of the Texas island chain was a group of shoals or offshore sandbars. Sandbars are affected by the two natural mechanisms that build barrier islands: winds and waves. Waves, breaking on the sandbars, deposited sand and shells. After thousands of years, the bars emerged as islands in the surf. Today, waves continue to add new material (accretion) and wash away material (erosion) from the islands, and wind-blown sand is deposited as dunes. But if not stabilized, dunes, like beaches, will erode, too. The system is extremely sensitive to outside forces and is constantly changing—sometimes slightly, sometimes dramatically. Barrier islands are home to a variety of plants and animals that live among their beaches, dunes, and grassland flats.

BARRIER ISLAND BEACHES

The beach habitat, probably the most familiar part of barrier islands to seaside vacationers, may appear barren. Waves rush onto bare sand in a seemingly endless, rhythmic motion. But although they are mostly void of vegetation, these beaches are still home to many creatures. Small crustaceans nestle in wet sand along the swash zone, where foaming water rushes up onto the beach as waves break. Microscopic plants and animals live in the shallow water and sand near the edge of the shore, too, but the most familiar plant life on the beach is found away from the waves, on the sand dunes.

Cockle shells washed up on a barrier island beach

BARRIER ISLAND DUNES

Directly behind the beach, sand dunes form a ridge along the front of the barrier islands. Sometimes bare, but often cloaked in a green mat of vegetation, dunes are formed by eolian (wind) processes. These same forces can obliterate the dunes they create. If a dune is not protected by vegetation, it can vanish as winds move the accumulated sand to other places on the island or offshore.

Vegetation slows this erosion. However, only a few species of plants can help the dunes, because only a few can live in the stressful environment. Many plants cannot tolerate the frequent winds, shifting sand, or salt spray from Gulf waters. The sandy substrate may be extremely dry, too dry for most plants.

The plants that do thrive in the dunes send out far-reaching roots that allow them to retrieve vital nutrients and moisture from the sand. At the same time, these roots trap sand and hold it in place. Their long, slender leaves, in which food production (photosynthesis) occurs, allow the plants to survive even when they

Barrier island dunes

are partially buried under shifting sands. These leaves are brittle and easily destroyed if trampled.

The most beautiful of these species is the tall, slender sea oats grass. Like other tall species, sea oats acts as a gentle windbreak, slowing breezes and causing them to deposit their load of sand. Other grasses, such as seacoast bluestem, coastal dropseed, paspalum, and marshhay cordgrass, also inhabit the dunes, although they are not as conspicuous or as stately as sea oats.

Creeping (prostrate) plants send out long runners which quickly form a lush green cover to bind loose, unconsolidated sand. These species take root and spread rapidly. They also add brilliant colors to the dunes with their flowers. The fiddleleaf morning glory produces a bright pink, funnel-shaped bloom, and the beach morning glory has white flowers

Roots holding dune sand in place

Sea oats

coast bluestem

Marshhay cordgrass

Coastal dropseed

Fiddleleaf morning glory

Beach morning glory

Sea ox-eye

Sunflower

Beach evening primrose

Beach tea

with yellow-tinted centers. Members of the sunflower family also are part of the dune floral community. The summer-blooming sea ox-eye and common sunflower stand out among the sea oats with their yellow, daisy-like flowers. The beach evening primrose is also abundant throughout the summer, but flowers only from early evening to late morning. The yellow blooms quickly wither and take on a reddish cast under the heat of the afternoon sun. Even small shrubs, such as beach tea, which is particularly abundant on Padre Island, have a place on the dunes.

Padre and Mustang islands both have excellent dune formations. Dunes on these southernmost islands are well developed, reaching heights of fifteen feet or more.

BARRIER ISLAND GRASSLAND FLATS

Behind the dunes, covering the central part of the islands, are extensive flat areas vegetated mainly by grasses. These are the barrier flats. They develop on deflation flats, areas that at one time were covered by dunes which have since been blown away; on washover fans, where sediment was laid down in a fan-shaped deposit by waves which breached the island; or, in some cases, in troughs in well-developed ridge-and-swale topography. Deflation flats, washover fans, and troughs are relict features, landforms produced by processes no longer operating there. What remains in each case is the same feature: a broad, flat place that is home to a greater variety of grasses than are found on the dunes.

Perhaps the most common grass in these areas is the seacoast bluestem, but Gulf cordgrass, Indiangrass, windmill grass, and paspalum grasses are also common. Another component of this community is a group of plants known as graminoids, or grasslike plants. Graminoids belong to two different families, the rushes and sedges, but are often confused with grasses.

Rushes colonize ponds or ditches in the flats. Like grasses, they have small, almost imperceptible flowers, but the lack of nodes (joints) in their round, hollow stems helps to distinguish them.

Sedges are more easily identified by their three-cornered or triangular stem. Their large but slender leaves often fold around the stem, radiating outwards. Sedges also have bracts, leaflike structures which surround the tiny flowering structures at the top of the plant. Sedges do not have large flowers with colorful petals. Their flowers are minute and usually white, beige, or a muted color. A hand

Barrier island grassland flats

lens or microscope can be used to see the intricacies of these tiny flowers.

An especially colorful and interesting sedge is the white-topped umbrella sedge, abundant on Padre Island. It has a very slender stem and cream-colored flowers grouped in a small cluster at the top of the plant. The bracts enclosing the flower cluster are unique because they are half white and half green. Most other sedges have bright green bracts.

The barrier flats are also home to goldenrods, with their gold-colored flowers clustered along tall stalks, and to the tiny, white saline aster, which looks like a miniature daisy. A few cacti are scattered throughout the flats, too. The prickly pear can grow in almost any dry environment and frequently appears in the flats as well as on the dunes. Because its flowers are large, showy, and colorful, it is easy to recognize.

White-topped umbrella sedge

Prickly pear

When rain comes to the barrier flats, it fills in depressed areas, creating small, ephemeral (short-lived) freshwater ponds. These ponds nourish different plants than those supported by the larger surrounding saline grassland ecosystem. Different plants invade and colonize these ponds, forming pockets of freshwater marshes. They are easily spotted, even from the roadside, because tall, slender cattails are prominent among the freshwater plants.

The large, brown, fuzzy spikes at the top of cattails are the plants' inflorescences, or flower clusters. Cattail flowers, like those of grasses and sedges, are tiny. Break the brown tip open, and you'll find wispy flowers that quickly scatter in the wind. Each is surrounded by a small tuft of hair that enables it to float in the seaside breezes.

Though cattails usually establish pure stands (an area dominated by one species) other plants can be found with them, including pennyworts, rushes, and spikerushes with tiny, slender, leafless stalks.

Cattail

Pennywort

Bays and Marshes

Texas could be considered to have two coastlines—one the barrier islands and the other the shoreline along the wetlands and bays found between the barrier islands and the mainland. The waters sandwiched between the two landforms are protected by the barrier chain from Gulf storms and daily wave action. They are relatively tranquil. But next to the mainland, bays and marshes, not sandy beaches, form the face between water and land.

Bays were formed during the Great Ice Age when lowerings of the sea level forced rivers to travel farther to deposit sediment in the Gulf. When sea levels dropped, the rivers' trips were longer; when sea levels rose, the deeply scoured river valleys flooded. These flooded river valleys or estuaries are today's bays. Each is surrounded by low-lying land that is frequently flooded by tides, rivers, or bayous. Because these lands are characterized by the constant presence of water, they are called wetlands. The wetlands (marshes), together with the bays (estuaries), are highly productive ecosystems, among the most biologically productive environments on the globe.

The bays and adjoining marshes, whether salt, brackish (part salt and part fresh water), or fresh, are home to the very beginnings of the food chain. The first links—plants—are present in the bays and marshes. Microscopic plants, phytoplankton, are eaten by microscopic animals, zooplankton, while the larger marsh plants are food for larger animals. Microscopic scavengers consume detritus, the decayed remains of plant life. All of these animals are eaten by larger animals, which, when they die, are decom-

The wetland food web

posed by bacteria, replenishing the soil and water with nutrients. Plants absorb these nutrients and can be, in turn, a direct food source for animals. If not, the plants decay and become detritus, and so on.

Each organism, plant or animal, large or small, contributes to the total ecosystem; none is an entity to itself. At the base of this food web, producing the original food to fuel the whole system, are the plants. And, because marshes are so abundant in plant life, they also support an abundant animal population, although sometimes only on a seasonal basis. Meanwhile, the shallow, relatively calm bay waters offer permanent as well as temporary homes to many fish and shellfish.

BAYS

Sea grasses in bay waters, although not a direct food source for fish, offer young fish protection from predators. But, more importantly, these "grasses" decay and produce biomass for the food web.

Sea grasses are not really grasses at all, but are submerged aquatic plants present throughout the year. Sea grass beds are found in and beneath the water. Four main species grow in Texas bays: shoal grass, widgeon grass, turtle grass, and manatee grass. All are found in each bay, usually growing in the shallower parts of bays and varying in abundance with local conditions. Water temperature, turbidity (cloudiness), and salinity all strongly affect the growth and distribution of each species.

One of Texas' major bays is really a lagoon. Laguna Madre, between the

A salt marsh

mainland and Padre Island, is a hypersaline lagoon, in which salinity is often higher than that of open ocean waters. Ocean water salinity averages 35 parts per thousand (ppt), but Laguna Madre can reach salinities of 40 ppt or greater. Laguna Madre is generally more salty because it has little input from freshwater rivers and the climate is arid, causing the evaporation rate to exceed rainfall. In spite of the high salinity, all four species of sea grass can be found in Laguna Madre, although shoal grass is dominant in the lagoon's upper reaches.

MARSHES

Three types of marshes line Texas bays: salt marshes, brackish marshes, and fresh marshes. Salt marshes are found along bays which receive salt water from the Gulf as well as from tides that inundate them. Fresh marshes are nearer creeks or ponds where tides have little effect and floodwaters are mainly from rivers. The area between these two, which fluctuates between saline and fresh water, is called brackish and is influenced by both salt and freshwater systems. The salinity of the soil underlying these marshes (substrate) depends on the water source, and the nature of both water and substrate affect the plant communities in each of the different marshes. All are subjected to frequent, if not constant, flooding and support different types of grasses, herbs, shrubs, and, less often, trees.

Salt Marshes. Salt marshes are common along the coast, fringing bays, river deltas, and the backside of barrier islands. Salt marsh vegetation is halophytic, or salt-loving. Salt water tends to dry out plant tissue, causing severe stress. But halophytes possess adaptations enabling them to colonize the salt marsh environment. Relatively few plants are halophytes, and by far the most interesting of these is the black mangrove.

Mangroves are usually tropical shrubs or trees that prefer a slightly warmer climate than Texas offers. However, the black mangrove has adapted to Texas and is the only mangrove established in the state. It can be found along the coast as far north as Galveston, with the healthiest and most robust stands near Port Aransas and South Padre Island. Dense stands form on marshy islands, providing resting sites for waterfowl as well as vital nutrients for the detrital cycle. This species is a pioneer of the salt marsh, colonizing wetter areas where few other plants can live. Its reproductive strategy is one reason it has been successful.

Mangroves are viviparous; their seeds germinate while still attached to the parent plant and are not released until they have developed into seedlings. When they fall from the parent plant, the seedlings immediately become fixed in the nearby substrate (if the water level is sufficiently low), or float in sea water un-

Black mangrove

til they reach substrate farther along the coast.

But although sea waters help the black mangrove by distributing seedlings, the water also floods the soils where mangroves grow, making oxygen transfer difficult. If soil is flooded, oxygen cannot get from the atmosphere to the soil, and the plants must find a way to get oxygen directly from the air. Black mangroves have developed specialized structures called pneumatophores to do just that.

Pneumatophores spring up in the soil surrounding the mangrove and ap-

Cross-section of Marshes

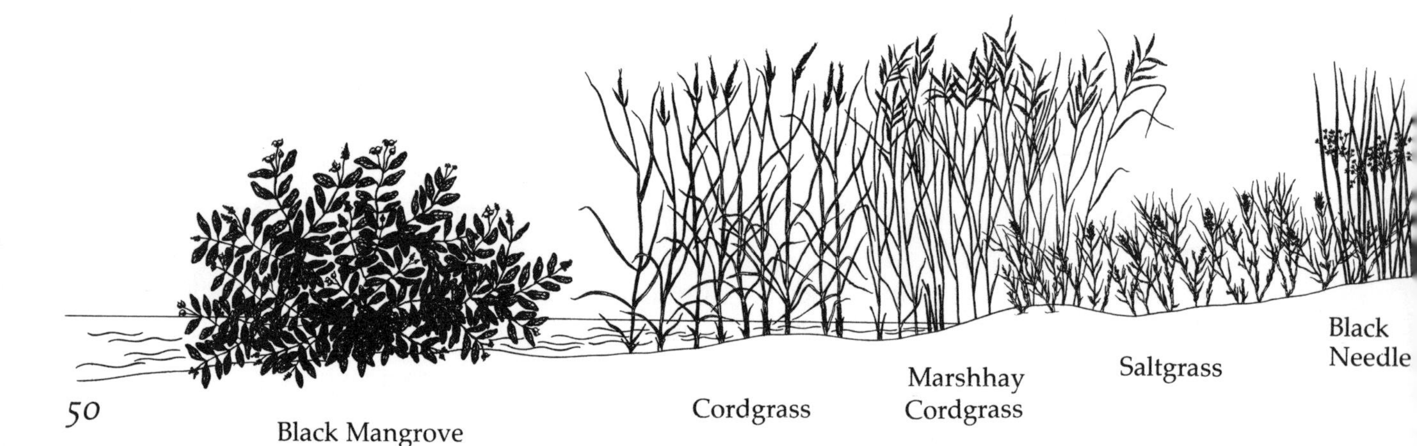

pear to be shoots of the closest plant. They are, in reality, aerating organs made of a specialized tissue (aerenchyma) that is spongy and airy and allows direct transfer of gases between plant and atmosphere. Pneumatophores allow the plant to "breathe" in its flooded environment.

Both pneumatophores and viviparous seedlings distinguish the black mangrove from other plants of the salt marsh, which do not have such exotic structures yet are able to cope adequately with the stressful living conditions. Long leaves of salt marsh grasses have special cells

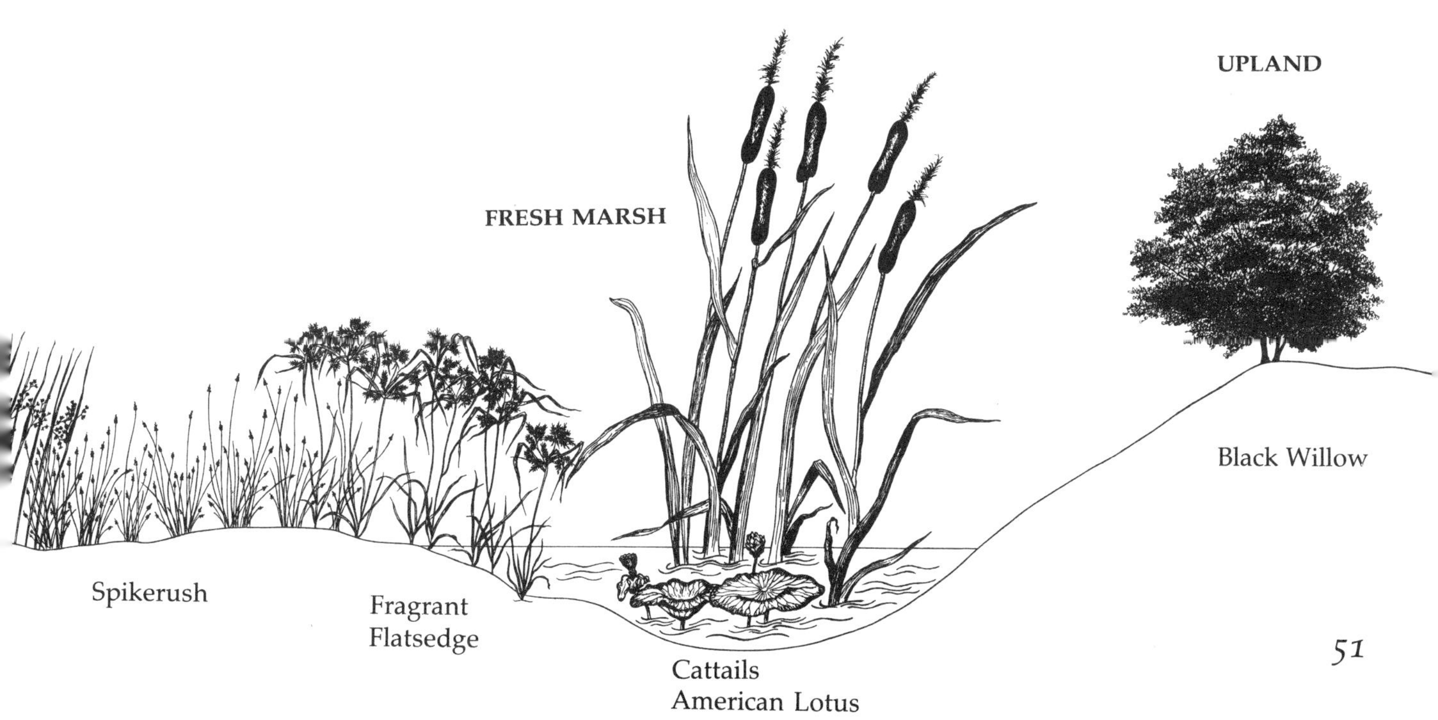

Black mangrove and saltwort

which allow oxygen from the atmosphere to be absorbed directly into those plants. One of the most abundant grasses, and a major contributor of detritus to the estuarine food web, is cordgrass. It often forms large, dense stands and grows along the coastal edges of the salt marsh, often with black mangroves. Glasswort and saltwort are succulent herbs that form mats over sites with high salinity. Saltgrass and marshhay cordgrass occupy the landward margins of salt marshes and invade brackish sites. More striking salt marsh flora include the sea ox-eye (also on barrier dunes), the marsh fleabane, with its small but numerous pinkish purple flowers, and the vividly tinted rose gentian.

Brackish Marshes. Other marshes landward of salt marshes are also subjected to frequent, though not as intense, flooding. These are the brackish marshes.

Cordgrass in a salt marsh

Glasswort

Glasswort and saltgrass

Sea ox-eye

Marsh fleabane

Since both tide and fresh waters flood brackish marshes, these marshes are neither saline nor fresh, but transitional. Vegetation there is more diversified than in salt marshes, although many plants in brackish marshes are also found in one of the other two marsh types. Marshhay cordgrass, so common in the upper levels of the salt marsh, is one of the most common elements of brackish marshes. Saltgrass and marshhay cordgrass form a dull green cover of thin, wiry grasses and are an important source of detritus for the estuarine food web.

Black needlerush, another source of detritus, is able to tolerate a wide range of salinities. It grows in salt marshes but reaches greater abundance in brackish marshes. In pure stands the plants have a black cast to them, but they are actually dark green. Black needlerush is one of the

Marshhay cordgrass

Black needlerush

Seaside heliotrope

most common rushes in Gulf coast marshlands.

Spikerush, a sedge, is another common wetland plant. The stems and seeds of these slender, leafless plants are an important food source for muskrats and various waterfowl.

The seaside heliotrope, whose seeds are consumed primarily by ducks, grows close to the ground in brackish marshes. These plants, scarcely noticeable except in sparsely populated areas, have delicate flowers with an unusual flowering stalk. Tiny white blooms appear along only one side of the stalk, which curls tightly at the end. Seaside heliotrope is also found on the seaward side of the dunes.

Fresh Marshes. Fresh marshes are those farthest from the sea, fed by creeks and bayous that run through the lowlands and drain into the bays. Since the water source there is so much different from that of the salt marshes, environmental conditions are different, and so are the plants.

Much of the fresh marsh vegetation, as in the other marshes, belongs to the graminoid families. The fragrant flatsedge, with its triangular stem, displays golden flower clusters that radiate outward from the main stem in a starlike pattern. Jamaican sawgrass, which is not a grass at all, but one of the largest sedges, has narrow, serrate (saw-edged) leaves and a large seed head, colored rust or brown. The soft-stem rush is light green and grows in dense clusters.

Two grasses common to fresh marshes are the Roseau cane, or common reed, and the bushy bluestem. Roseau cane is a very tall grass, sometimes reaching twelve feet in height. With its large plumelike inflorescence, it resembles sugar cane. Dense stands of Roseau cane

Fragrant flatsedge

provide resting and breeding grounds for various forms of wildlife. Bushy bluestem, usually found at slightly elevated sites, is easily recognized by its bushy, broomlike head. It is shorter than Roseau cane, usually two to four feet tall, and somewhat more prevalent in Texas marshes. It also harbors wildlife.

Many other marsh plants are not so easily recognized because they are much smaller than the grasses and graminoids. The yellow deer-pea, a vine with bright green leaves and small but numerous yellow flowers, twines itself around most other plants, sometimes forming mats over them and providing browse for deer. The arrowhead is an easily recognized species because it has small, three-petaled white flowers and large, bright green, arrow-shaped leaves. Flowers of the bulltongue, a related species, are similar, but its leaves are lance-shaped.

Not all marsh plants are useful; some are detrimental to the ecosystem. The most well known of these is the water hyacinth, originally brought into this

Roseau cane

Yellow deer-pea

country as an ornamental. It managed to escape cultivation and is seen today in roadside ditches, bayous, and other slow-moving waterways, forming a stunning lavender and green mat that chokes and clogs the channels. Although the leaves are eaten by deer, nutria, and other herbivores, the damage they do by clogging waterways, impeding fishing, and shading out more beneficial food plants greatly

Water hyacinth

American lotus

outweighs their usefulness as forage. They spread rapidly and are incredibly difficult to control or eliminate.

Equally beautiful, but less destructive, is the American lotus. Its large, yellow flowers can be seen in shallow freshwater ponds. The large, round leaves seem to float on top of the ponds, yet are rooted to the bottom by long, slender leafstalks (petioles). Though lotus tends to grow in slightly scattered stands and does not threaten waterways as does the water hyacinth, it does shade out other plants more valuable as wildlife food.

Chenier Plain

The chenier plain of eastern Texas is unique. It lacks both bays and a barrier island buffer. Instead, this eighty-mile stretch of coast is edged by a small, narrow strip of beach. Behind the beach are flat marshes divided by elevated ridges that parallel the coast. These strips of elevated land are the cheniers (from the French word *chene*, meaning "oak," the natural vegetation on the ridges). Because they are topographic highs, the cheniers are prime land for residential development, agricultural development, or pasture.

The chenier plain is a very young coastal feature, created by the Mississippi River. The Mississippi drains an extensive area, carrying a tremendous load of mud, sand, and other sediments. Once deposited at the river's mouth, this material is carried along the coast by the alongshore

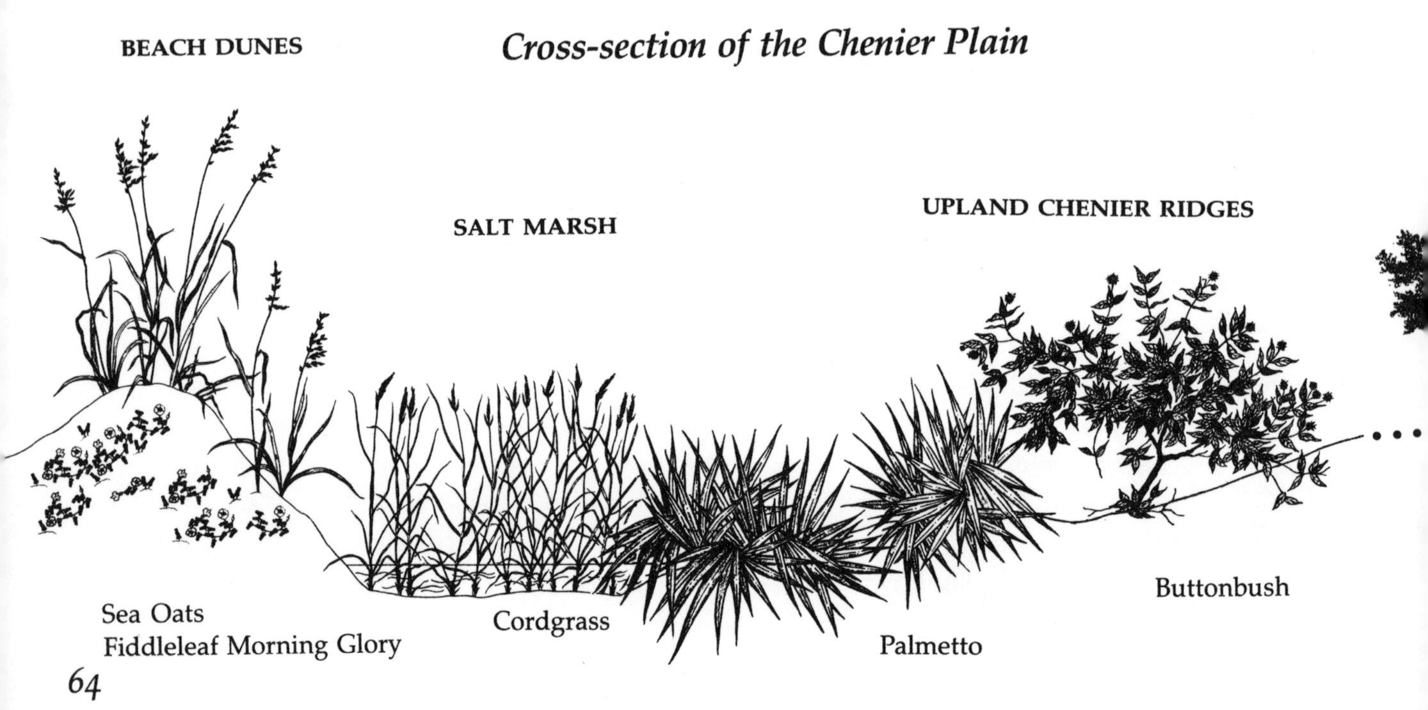

Cross-section of the Chenier Plain

current, which moves in a generally westward direction. The mud settles along the Gulf coast, forming low-lying marshy areas. However, the Mississippi shifts channels often. When the river flows in a western lobe of its delta, sediment is carried to the Texas shore. When it empties through an eastern lobe, as it does today, the sediment supply is cut off. At those times the Texas shore gets sand and shell

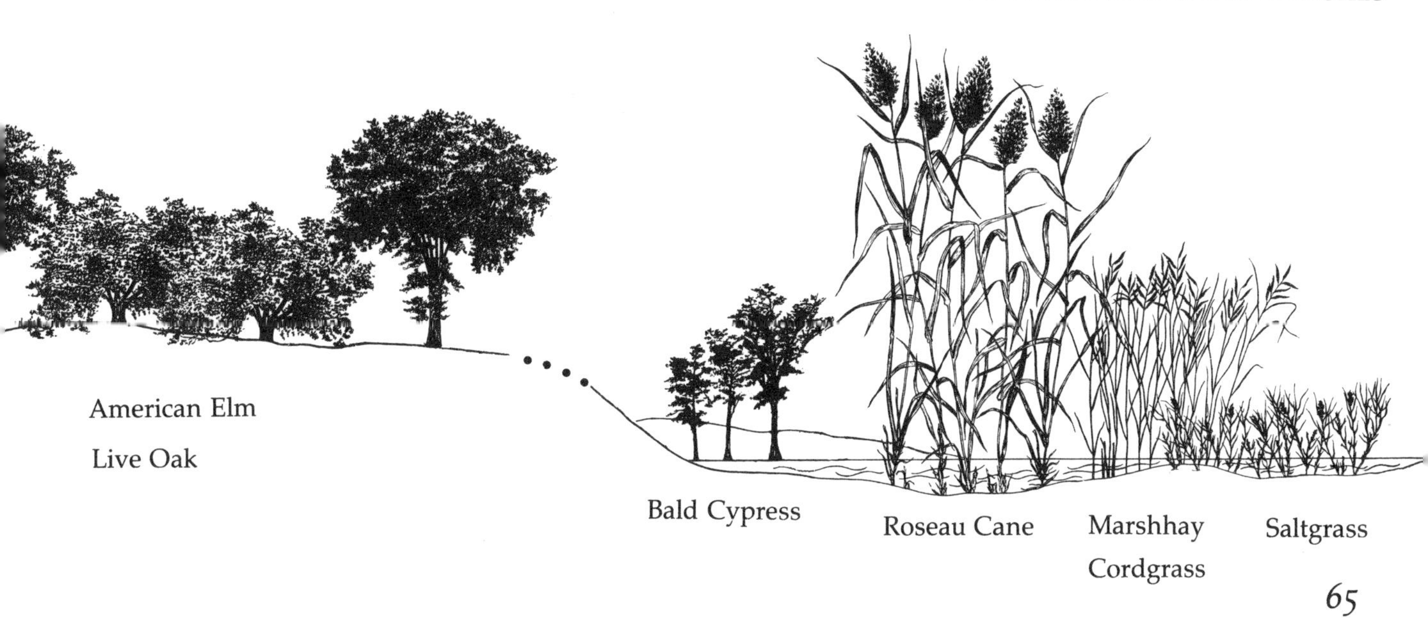

from offshore, brought in by waves and storms. Through time this sand and shell deposition builds a series of ridges—the cheniers of southwestern Louisiana and eastern Texas—that lie parallel to the coast and are separated by marshes or swales.

CHENIER BEACHES

Chenier beaches are similar to other beaches along the coast. Though not as large as beaches on the barrier islands, chenier beaches, such as those found at Sea Rim State Park, offer an abundance of shells. Stretches of the beach have been covered with shells washed ashore by hurricanes, which commonly obliterate dunes and highways.

Moonsnail and other shells

But not all the chenier plain dunes have been eroded. Dune fields exist along Sea Rim State Park, although they are not as large as those in South Texas. Vegetation is similar to that on dunes in other parts of the coast, with sea oats, sea ox-eye, and fiddleleaf and beach morning glories prominent. Additional species are found on these dunes. The showiest is the rosy marsh mallow, usually found in fresh to brackish marshes. Its bright pink flowers, long, yellow, fused stamens, and toothed leaves make the mallow noticeable among the grasses and asters of the dunes.

Fiddleleaf morning glory

Rosy marsh mallow

CHENIER MARSHES

Backing the chenier dunes and beaches are marshes—the low-lying features we saw earlier near the bays. These marshes do not surround bays but lie parallel to the coast in between the chenier ridges. They are usually either salty or brackish, depending on how close they are to the sea and how often they are flooded. Most plants common to these marshes are similar to marsh plants elsewhere along the coast. Salt marshes are often inhabited by nearly pure stands of cordgrass, with a few sedges interspersed, especially near streams within the marsh. No mangroves are found in these salt marshes, however, because frost is common there in the winter, and those plants cannot survive the cold. In the brackish marshes, Roseau cane is found in large stands, along with salt grass and marshhay cordgrass. Along el-

Roseau cane Cordgrass

evated hummocks (small hills or mounds) in the marsh are shrubs such as the marsh elder and groundsel tree and herbs such as the yellow deer-pea and saltmarsh asters.

CHENIER RIDGES

The chenier or ridge habitat is unique along the coast because it provides topographic relief up to ten feet above mean sea level in an otherwise flat landscape. Because they are above normal tidal influence, cheniers support a variety of vegetation types, including live oaks and other hardwood trees. Their elevation also makes cheniers attractive for extensive use by man; consequently, few of the oaks remain. Vegetation in these areas includes common upland species such as hackberry, American elm, red maple, and some wetland species, including bald cypress and black willow. Palmettos, but-

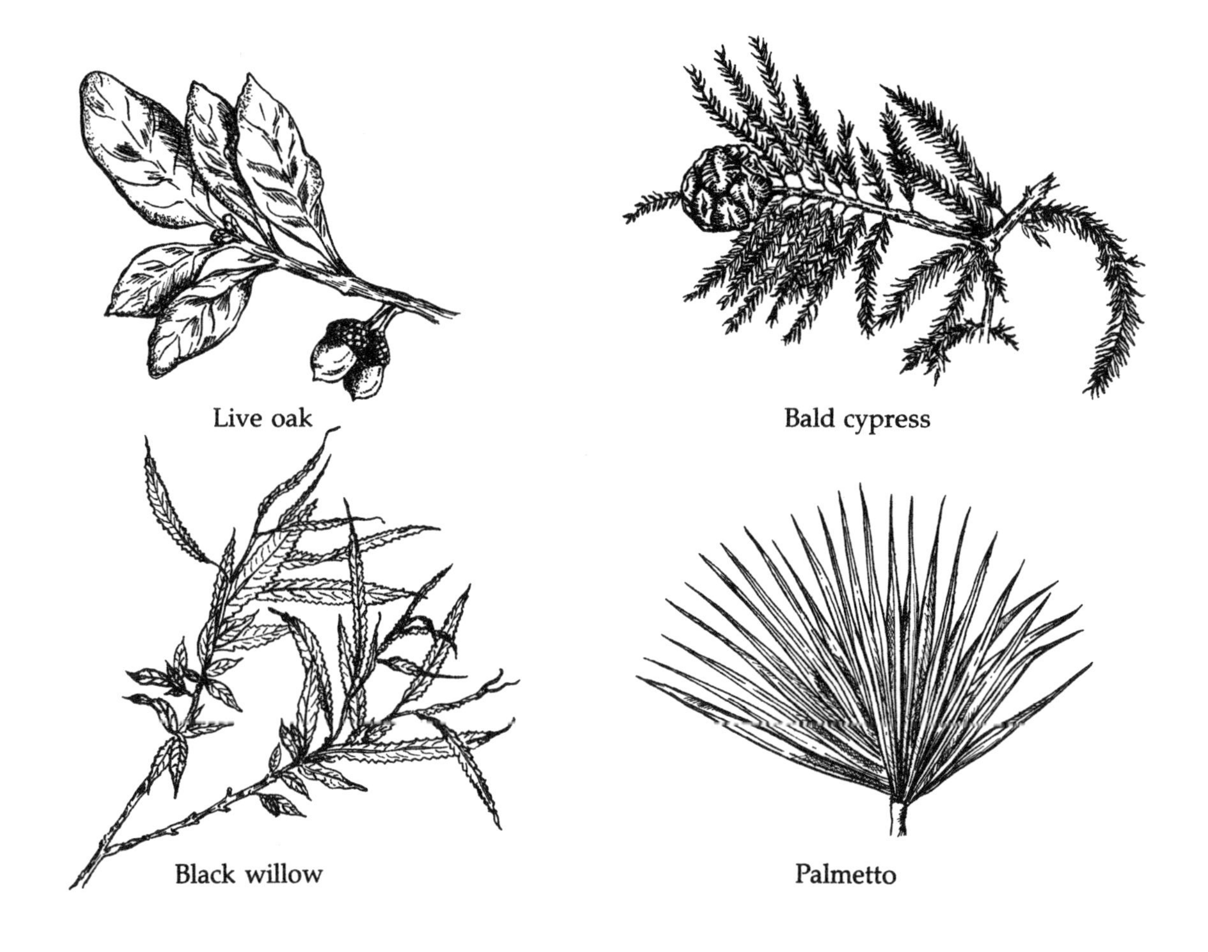
Live oak
Bald cypress
Black willow
Palmetto

A chenier ridge at water's edge

tonbush, and marshhay cordgrass grow among the trees.

Across the Coastal Zone

All along the coastal zone, parallel to stream courses, are riverbanks created by occasional flooding after heavy rains. Built of sand, silt, and other alluvium, these banks support a rich flora of tree species—bay, pecan, oak, and elm—as well as vines (muscat grapes and cat's-claw greenbrier) and herbaceous plants.

Some Additional Information

If you are planning a trip to the Texas coast, feel free to request information on lodging, historical points of interest, museums, or other recreational facilities from the chambers of commerce in coastal cities. Refuges and parks in the coastal zone include the following:

Sea Rim State Park. On Highway 87 about twenty miles southwest of Port Arthur, Sea Rim consists of both a marsh and a beach unit. Guided airboat tours through the marsh are available March through October. For more information, contact Sea Rim Airboat Service, P.O. Box 517, Sabine Pass, Texas 77655. Telephone: (409) 971-2963.

Anahuac National Wildlife Refuge. On East Bay, east of Galveston and north of High Island, this refuge was established and is managed primarily for migrating and wintering waterfowl. For information on accessibility to the refuge, contact Refuge Manager, Anahuac National Wildlife Refuge, P.O. Box 278, Anahuac, Texas 77514. Telephone: (409) 267-3337.

Aransas National Wildlife Refuge. Thirty-five miles northeast of Rockport, via Texas 35 north, Aransas is famed as a principal wintering ground for whooping cranes. The Wildlife Interpretive Center features mounted specimens, slide shows, and literature. The greatest number of species is at the refuge between November and March. For more information, contact Refuge Manager, Aransas National Wildlife Refuge, Box 100, Austwell, Texas 77950. Telephone: (512) 286-3559.

Welder Wildlife Refuge. Eight miles northeast of Sinton, on U.S. Highway 77, this refuge was created by provisions made in rancher Bob Welder's will. The diversity of vegetation there provides outstanding habitats for both wildlife and cattle. The Welder Wildlife Foundation provides fellowships for graduate students interested in wildlife ecology and management. Public tours are held on Thursdays, and group or special tours may be requested by contacting Welder Wildlife Refuge, Drawer 1400, Sinton, Texas 78387, or the Tour Director, (512) 364-2643.

Brazoria/San Bernard Wildlife Refuge. Near Angleton, this refuge is a major goose wintering area. Major habitats include salt marsh interspersed with brackish lakes and ponds. For more information, contact Refuge Manager, Brazoria/San Bernard National Wildlife Refuge, Box 1088, Angleton, Texas 77515. Telephone: (409) 849-6062.

Padre Island National Seashore. Occupying over 80 miles of the 110-mile-long Padre Island, this seashore stretches from Corpus Christi southward to near Brownsville. Though most of the seashore remains in its natural state, the federal government has provided a complex of recreation facilities near the island's northern end, including a visitor and interpretive center. For more information, contact Superintendent, Padre Island National Seashore, 9405 South Padre Island Drive, Corpus Christi, Texas 78418. Telephone: (512) 937-2621.

Laguna Atascosa National Wildlife Refuge. Twenty-five miles east of Harlingen in extreme southern Texas, this refuge encompasses about 46,000 acres, including 7,000 acres of open water and marshes. It is a winter haven for ducks and geese as well as many other forms of wildlife. For visitor information, contact Refuge Manager, Box 450, Rio Hondo, Texas 78583. Telephone: (512) 748-3607.

Plants Listed By Habitat As Noted In Text

Habitat	Common Name	Scientific Name
Dunes	Beach evening primrose	*Oenothera drummondii*
(barrier island and	Beach morning glory	*Ipomoea stolonifera*
chenier plain)	Beach tea	*Croton punctatus*
	Coastal dropseed	*Sporobolus virginicus*
	Common sunflower	*Helianthus* spp.
	Fiddleleaf morning glory	*Ipomoea pes-caprae*
	Marshhay cordgrass	*Spartina patens*
	Paspalum	*Paspalum* spp.
	Rosy marsh mallow	*Kosteletzkya virginica*
	Sea oats	*Uniola paniculata*
	Sea ox-eye	*Borrichia frutescens*

Nomenclature follows that of D. S. Correll and M. C. Johnston, *Manual of the Vascular Plants of Texas* (Renner: Texas Research Foundation, 1970), except for the grasses, for which F. W. Gould, *The Grasses of Texas* (College Station: Texas A&M University Press, 1975), has been used.

Habitat	Common Name	Scientific Name
	Seacoast bluestem	*Schizachyrium scoparium*
Grassland Flats (*Freshwater ponds within the flats)	Cattails*	*Typha* spp.
	Goldenrod	*Solidago* spp.
	Gulf cordgrass	*Spartina spartinae*
	Indiangrass	*Sorghastrum* sp.
	Paspalum	*Paspalum* spp.
	Pennywort*	*Hydrocotyle* spp.
	Prickly pear	*Opuntia lindheimeri*
	Rush*	*Juncus* spp.
	Saline aster	*Aster subulatus*
	Seacoast bluestem	*Schizachyrium scoparium*
	Spikerush*	*Eleocharis* spp.
	White-topped umbrella sedge	*Dichromena colorata*
	Windmill grass	*Chloris* sp.
Bays	Manatee grass	*Cymodocea filiformis*
	Shoal grass	*Halodule beaudettei*
	Turtle grass	*Thalassia testudinum*
	Widgeon grass	*Ruppia maritima*

Salt Marshes (including chenier plain)	Black mangrove	*Avicennia germinans*
	Cordgrass	*Spartina alterniflora*
	Glasswort	*Salicornia* spp.
	Groundsel tree	*Baccharis halimifolia*
	Marsh elder	*Iva frutescens*
	Marsh fleabane	*Pluchea purpurascens*
	Marshhay cordgrass	*Spartina patens*
	Rose gentian	*Sabatia* sp.
	Saltgrass	*Distichlis spicata*
	Saltwort	*Batis maritima*
	Sea ox-eye	*Borrichia frutescens*
Brackish Marsh (including chenier plain)	Black needlerush	*Juncus roemerianus*
	Marshhay cordgrass	*Spartina patens*
	Saltgrass	*Distichlis spicata*
	Saltmarsh aster	*Aster subulatus*
	Seaside heliotrope	*Heliotropium curassavicum*
	Spikerush	*Eleocharis* spp.
	Yellow deer-pea	*Vigna luteola*

Fresh Marshes	American lotus	*Nelumbo lutea*
	Arrowhead	*Sagittaria latifolia*
	Bulltongue	*Sagittaria lancifolia*
	Bushy bluestem	*Andropogon glomeratus*
	Fragrant flatsedge	*Cyperus odoratus*
	Jamaican sawgrass	*Cladium jamaicense*
	Knotweed	*Polygonum* spp.
	Roseau cane	*Phragmites australis*
	Soft-stem rush	*Juncus effusus*
	Water hyacinth	*Eichhornia crassipes*
	Yellow deer-pea	*Vigna luteola*
Chenier Ridges	American elm	*Ulmus americana*
	Bald cypress	*Taxodium distichum*
	Black willow	*Salix nigra*
	Buttonbush	*Cephalanthus occidentalis*
	Hackberry	*Celtis laevigata*
	Live oak	*Quercus virginiana*
	Palmetto	*Sabal minor*
	Red maple	*Acer rubrum*
	Marshhay cordgrass	*Spartina patens*